Tons of Trucks!

FORKLIFTS

By Luke Harasymiw

Gareth Stevens
PUBLISHING

Please visit our website, www.garethstevens.com. For a free color catalog of all our high-quality books, call toll free 1-800-542-2595 or fax 1-877-542-2596.

Library of Congress Cataloging-in-Publication Data

Names: Harasymiw, Luke, author.
Title: Forklifts / Luke Harasymiw.
Description: New York : Gareth Stevens Publishing, [2020] | Series: Tons of trucks! | Includes index.
Identifiers: LCCN 2019010241| ISBN 9781538245095 (pbk.) | ISBN 9781538245118 (library bound) | ISBN 9781538245101 (6 pack)
Subjects: LCSH: Forklift trucks–Juvenile literature.
Classification: LCC TL296 .H375 2020 | DDC 621.8/63–dc23
LC record available at https://lccn.loc.gov/2019010241

First Edition

Published in 2020 by
Gareth Stevens Publishing
111 East 14th Street, Suite 349
New York, NY 10003

Editor: Therese Shea
Designer: Katelyn E. Reynolds

Photo credits: Cover, p. 1 K.D.P/Shutterstock.com; p. 5 JRP Studio/Shutterstock.com; p. 7 Dmitry Kalinovsky/Shutterstock.com; p. 9 Salvador Aznar/Shutterstock.com; p. 11 Pissanu Jirakranjanakul/Shutterstock.com; pp. 13, 17 Halfpoint/Shutterstock.com; pp. 15, 24 (fork) Pressmaster/Shutterstock.com; p. 19 Imfoto/Shutterstock.com; pp. 21, 24 (batteries) Corepics VOF/Shutterstock.com; p. 23 People Image Studio/Shutterstock.com.

Printed in the United States of America

CPSIA compliance information: Batch #CW20GS: For further information contact Gareth Stevens, New York, New York at 1-800-542-2595.

Contents

This is my toy truck.
It's a forklift!

This is a real forklift.
It's a useful truck!

It lifts heavy loads.

EXTINTOR DE INCENDIOS
2SDD
HYSTER1.75

Most are about the size of a car.

C
D
G
35
9

Some lift loads high!

The fork is on the front.
The fork lifts the load.

13
3
14
7
15
8

A weight is on the back.
The truck won't tip.

This forklift carries boats!

Raytheon

Some use batteries!
They work for hours.

These trucks help!

Words to Know

batteries

fork

Index